Laura Imperatori

Non-Linearity. Frequency-doubling, degree variation, amplitudemodulation and demodulation

GRIN Verlag

Bibliografische Information der Deutschen Nationalbibliothek:

Die Deutsche Bibliothek verzeichnet diese Publikation in der Deutschen National-bibliografie; detaillierte bibliografische Daten sind im Internet über http://dnb.d-nb.de/ abrufbar.

Imprint:

Copyright © 2013 GRIN Verlag GmbH
Druck und Bindung: Books on Demand GmbH, Norderstedt Germany
ISBN: 978-3-656-66760-5

This book at GRIN:

http://www.grin.com/en/e-book/266748/non-linearity-frequency-doubling-degree-variation-amplitudemodulation

GRIN - Your knowledge has value

Der GRIN Verlag publiziert seit 1998 wissenschaftliche Arbeiten von Studenten, Hochschullehrern und anderen Akademikern als eBook und gedrucktes Buch. Die Verlagswebsite www.grin.com ist die ideale Plattform zur Veröffentlichung von Hausarbeiten, Abschlussarbeiten, wissenschaftlichen Aufsätzen, Dissertationen und Fachbüchern.

Visit us on the internet:

http://www.grin.com/

http://www.facebook.com/grincom

http://www.twitter.com/grin_com

NON-LINEARITY - FREQUENCY-DOUBLING, DEGREE VARIATION, AMPLITUDE MODULATION AND DEMODULATION

LAURA IMPERATORI

MURRAY EDWARDS COLLEGE

LSI22

PRACTICAL PARTNER: GOODWIN GIBBINS

EXPERIMENT PERFORMED THURSDAY, OCTOBER 25, 2012

ABSTRACT. In order to test non-linearity, the effects of different transfer functions of an AD633 multiplier in a given electrical circuit were investigated and compared with the theoretical expectations. First of all, the phenomenon of frequency doubling was found to occur when squaring the input voltage. Secondly, the multiplier was reconfigured to give a square-root response. This allowed us to vary the degree of non-linearity by choosing the parameters of input voltage and DC offset such that we could determine which terms in the Taylor expansion of the transfer function were relevant and hence to what degree the circuit behaved non-linearly. For a small, sinusoidal variation about a large DC level, the system was found to be weakly non-linear. For high amplitude and a low DV offset we observed strong non-linearity. Compared to weak non-linearity, we were able to detect the third harmonic as well as the first and the second one. The existence of harmonics was investigated on the PicoScope screen and verified by plotting output amplitude (dBV) versus input amplitude (dBV) and finding the gradient of the slope corresponding to the respective harmonic. Finally, frequency mixing was explored in its broader context by investigating amplitude modulation and demodulation on the same circuit board.

1. INTRODUCTION

The aim of the experiment is to investigate the behaviour of non-linearity in electrical circuits. Most common circuit elements such as capacitors, inductors and resistances as well as operational amplifiers behave very linearly and thus enable an intuitive understanding of the qualitative behavior of the circuit, but most physical systems are inherently nonlinear in nature such as neural circuits [?] and climate dymamics [?]. Therefore, nonlinear problems are of interest to engineers, physicists and mathematicians. The most common consequence of non-linearity is frequency-doubling which is also known as Second Harmonic Generation. It is a nonlinear optical process, in which photons interacting with a nonlinear material are effectively "combined" to form new photons with twice the energy, and therefore twice the frequency and half the wavelength of the initial photons. It was first showed by Peter Franken et al. at the University of Michigan in 1961.[?]

2. THEORETICAL BACKGROUND

2.1. Non-linear circuits in constrast to linear circuits.
In linear circuits, the output voltage is a function of the input voltage: $V_{out} = LV_{in}$, where

$$(1) \qquad L(\alpha V_{in,1} + \beta V_{in,2}) = \alpha V_{out,1} + \beta V_{out,2}$$

Date: September 18, 2013.

1

Hence, linear circuits obey the superposition principle. For a sinusoidal input voltage of frequency f, any steady-state output of the circuit (the current through any component, or the voltage between any two points) is also sinusoidal with frequency f. Examples of linear circuits are circuits composed exclusively of ideal resistors, capacitors, inductors, and operational amplifiers in the non-saturated regime, such as small-signal amplifiers, differentiators, and integrators. Some examples of nonlinear electronic components are: diodes, saturated transistors, iron core inductors and transformers. These non-linear components obey a transfer function, which specifies how the input voltage is modified. In this experiment, we introduced the AD633 multiplier, which takes two input voltages and multiplies them together giving an output voltage. Hence, its transfer function is as follows: $V_{out} = V_x \times V_y/10$.

2.2. Frequency doubling.

In the given experimental setup, frequency doubling can be explained by looking at the effect of the multiplier. For an input voltage $V_{in} = A\cos\omega t$, the output of the AD633 multiplier is

$$(2) \qquad W(t) = X(t)^2 = BA^2(\cos\omega t)^2 = BA^2\left(\frac{1+\cos 2\omega t}{2}\right)$$

Hence, there are two different frequencies in the output: a DC-offset and twice the frequency of the input voltage. The transfer function of a general non-linear system, $V_{out} = f(V_{in})$ can be approximated by a Taylor series, which with sinusoidal input $A\cos\omega t$, gives

$$(3) \qquad V_{out} = f(V_{in}) = \sum_n a_n V_{in}^n = \sum_n a_n A^n (\cos\omega t)^n$$

Using power-reduction formula, the $(\cos\omega t)^n$ terms can be written as sums of terms with frequencies that are multiples of ωt. Thus, a non-linear system generates an output with frequency components of $\omega, 2\omega, 3\omega, 4\omega$,etc.

2.3. Varying the degree of non-linearity.

The square-root circuit transforms the output voltage into the square-root of the input voltage: $V_{out} = \sqrt{-10V_{in}}$. This equals $\sqrt{10B + 10A\cos(\omega t)}$ on defining $-V_{in}$ as $B + A\cos(\omega t)$. Hence, the Taylor series of V_{out} can be obtained by multiplying the Taylor expansion of $f(x) = \sqrt{1+x}$ for $x = \frac{A}{B}\cos\omega t$ with the factor $\sqrt{10B}$:

$$
\begin{aligned}
(1+x)^{\frac{1}{2}} &= 1 + \frac{x}{2} - \frac{x^2}{8} + \frac{x^3}{16} - \frac{5x^4}{128} + \mathcal{O}(x^5) \\
&= 1 + \frac{A\cos(\omega t)}{2B} - \frac{A^2\cos(\omega t)^2}{8B^2} + \frac{A^3\cos(\omega t)^3}{16B^3} - \frac{5A^4\cos(\omega t)^4}{128B^4} + \mathcal{O}(x^5) \\
&= \frac{1}{1024B^4}[(-15A^4 - 64A^2B^2 + 1024)B^4 + 16AB(3A^2 + 32B^2)\cos(\omega t)
\end{aligned}
$$

$$(4)$$

$$
- 4A^2(5A^2 + 16B^2)\cos(2\omega t) + 16A^3B\cos(3\omega t) - 5A^4\cos(4\omega t)] + \mathcal{O}(\cos(\omega t)^5)
$$

This can be simplified to obtain the harmonic coefficients of the series $\cos(\omega t), \cos(2\omega t), \cos(3\omega t)$:

$$(5) \qquad c_0 = \frac{-15A^4 - 64A^2B^2 + 1024)B^4}{1024B^4}$$

$$(6) \qquad c_1 = \frac{16AB(3A^2 + 32B^2)}{1024B^4}$$

$$(7) \qquad c_2 = \frac{-4A^2(5A^2 + 16B^2)}{1024B^4}$$

$$(8) \qquad c_3 = \frac{16A^3B}{1024B^4}$$

$$(9) \qquad c_4 = \frac{-5A^4}{1024B^4}$$

Here we can see that $\cos(\omega t)^3$ contains a term $\cos(\omega t)$ as well as $\cos(\omega t)^4$ includes a term $\cos(2\omega t)$. These coefficients show that large values of the DC-offset (B) and small amplitudes of the input voltage (A) imply a small number of relevant terms in the Taylor series, which corresponds to weak non-linearity, whereas small values in B and large amplitude values correspond to big changes in the function of the output voltage, hence to strong non-linearity. In general, the smaller the gradient, the smaller amout of terms in the corresponding Taylor series.

2.4. **Amplitude modulation and demodulation.** Electromagnetic energy can only be transferred in an efficient way if the dipole length fits well to the wavelength of the corresponding frequency. An acoustic signal of frequency $f = 1kHz$ would have to be emitted by an antenna of length $l = \frac{c}{2f} = 150km$. As this is not very practical, signals are transformed to be emitted at a higher frequency via amplitude, frequency or phase modulation: A high carrier frequency is multiplied by the signal frequency. In the time domain, this gives rise to an envelope of the carrier frequency, which contains the information of the signal frequency. In the frequency domain, amplitude modulation produces a signal with power concentrated at the carrier frequency and two adjacent sidebands, each one equal in bandwidth to that of the modulating signal. The data is shown in Figure ?? and ?? respectively. This phenomenon can be explained mathematically by frequency mixing. If two frequencies ω_1 and ω_2 are fed into the multiplier, the output will be:

$$(10) \qquad A\cos(\omega_1 t) \times B\cos(\omega_2 t) = \frac{1}{2} \times AB[\cos((\omega_1 + \omega_2)t) + \cos((\omega_1 - \omega_2)t)]$$

The original signal can then be recovered from the modulated waveform by a further multiplication with the carrier.

$$\frac{1}{2} \times AB[\cos(\omega_1 + \omega_2 t) + \cos(\omega_1 - \omega_2 t)] \times A\cos(\omega_1 t)$$

$$(11) \qquad = \frac{1}{2} \times A^2B\left[\frac{\cos(2\omega_1 + \omega_2 t)}{2} + \frac{\cos(2\omega_1 - \omega_2 t)}{2} + \cos(\omega_2 t)\right]$$

3. METHODS AND RESULTS

The following measurements of non-linear behaviour in electrical circuits due to a non-linear compononent (AD633 multiplier) were conducted based on the Printed Circuit Board in Figure ??.

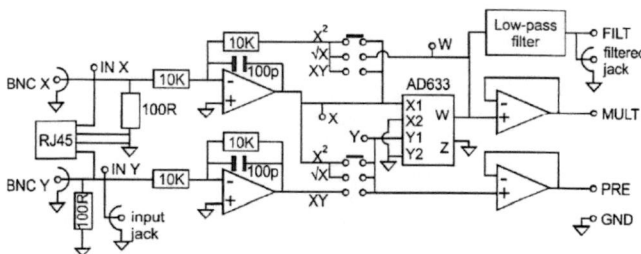

FIGURE 1. The main circuitboard with the non-linear component - the AD633 multiplier. Special features of this circuit are the two inverting amplifiers (-1 buffers) before as well as the two +1 buffers after the AD366.

3.1. **Frequency doubling.** In order to investigate a non-linear system with a response that depends on the square of the input, the AD366 multiplier was set to X^2. This then simplified the given circuit diagram in Figure ??.

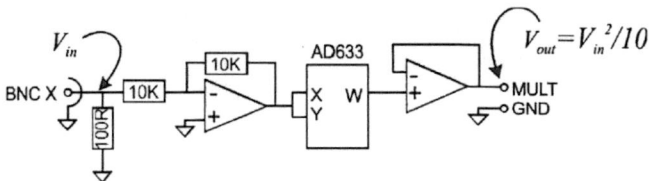

FIGURE 2. The key features are the -1 and +1 buffer.

The BNC X input was connected to the signal generator at the 50Ω output. The input signal was then observed at the test point IN X and the output was available at the test point MULT. The different frequency values were measured using the built-in function of the PicoScope Software. The different amplitudes were measured with the ruler function to get more precise values and to estimate errors better. At $100kHz$ there was no more a visible amplitude of the output signal. The gain of operational amplifiers decreases at high frequencies due to saftey reasons.[1] Considering the two operational amplifiers in the circuit, we decided to conduct the amplitude measurements at a frequency of $1.066kHz$. Based on our theoretical prediction, we expected a straight line with gradient 2 for the plot of the output

---

[1]Over a big enough frequency range, the feedback resistor which has inductive and capacitive components will lead to a phase shift of π. This will change negative feedback to posive feedback, which implies oscillation rather than amplification.

versus input frequency as well as for the log-log plot of output versus input voltage. The experimental results agreed very well with our expectations.

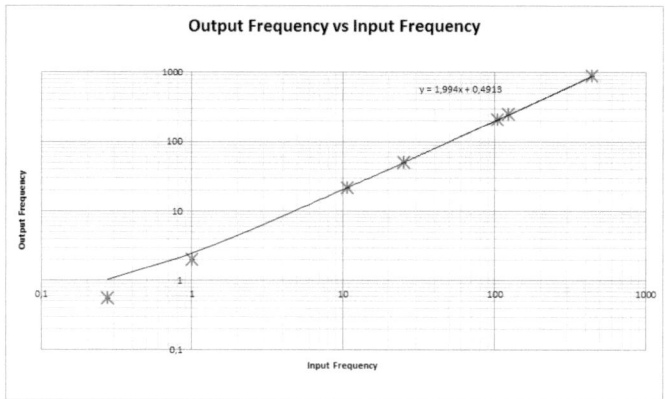

FIGURE 3. The trendline fit agrees very well within the range of errors. The value of the y-intercept is $\frac{BA^2}{2}$.

By plotting the output voltage versus the input voltage, we saw that $V_{out} = \frac{(V_{in})^2}{10}$, as the fitted trendline was approximately $\log V_{out} = 2 \log V_{in} - \log 10$.

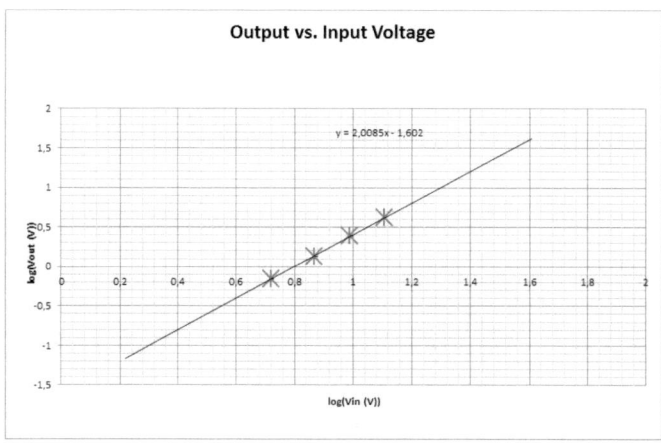

FIGURE 4. The fit of the trendline is very good with $R^2 = 0.9993$, but the y intercept is expected to be $-\log 10 = -1$.

3.2. **Square-root circuit.** The switch of the printed circuit board was switched to $\sqrt{(x)}$, which transforms the PCB into this circuit. This circuit was then used to investigate different degrees of non-linear behaviour. If we apply the Golden Rules to the inverting operational amplifier, we see that the value of W has to determined by the feedback components and thus has a value of $-V_{in}$ at low frequencies, even though the circuit is non-linear. The output amplifier then buffers the input of the multiplier which acts as a squarer. Comparing these two conditions on W, we see that $W = -V_{in} = \frac{V_{out}^2}{10}$, hence $V_{out} = \sqrt{-10V_{in}}$. In fact, W was measured to be proportional to $-V_{in}$. To test this squareroot-relationship of the transfer function, the data was collected via the inherent measurement function in the PicoScope Software that exported the data to Excel. 8000 data points were then collected, but only 1000 could be included in the Excel graph. This was sufficient data for our purposes as our aim was to verify the square-root relationship of the AD633

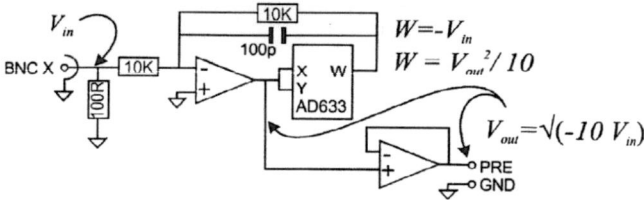

FIGURE 5. The main features are the multiplier within the feedback loop of the first inverting amplifier, the output amplifier with a gain of +1 that buffers into the AD633 as well as the capacitor, which stabilises the feedback loop.

By plotting the output voltage versus the input voltage, we saw that $V_{out} = A\sqrt{V_{in}}$, as the fitted trendline was $log(V_{out}) = log(A) + \frac{1}{2}log(V_{in})$.

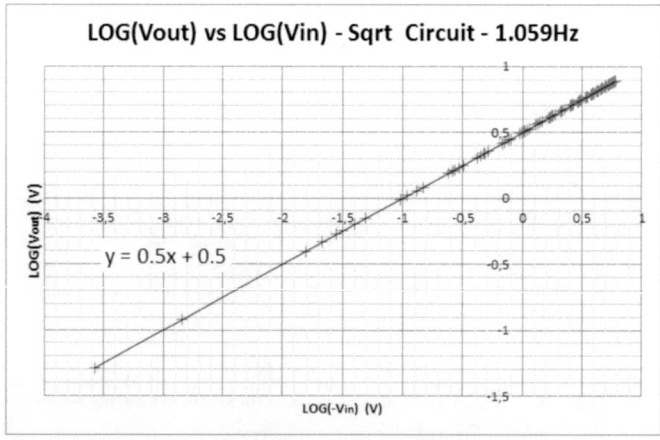

FIGURE 6. Log-Log plot to demonstrate that square-root relationship exists at test point PRE.

3.3. **Weak Non-Linearity.** The measurements were conducted in dbV, which implies that we plotted logarithmically by definition, as $L_{dB} = 20 \log \frac{V_{out}}{V_{ref}}$. We also observed that by increasing the DC-Offset, the second harmonic coefficient decreased as can be theoretically predicted from the Taylor expansion(**??**).

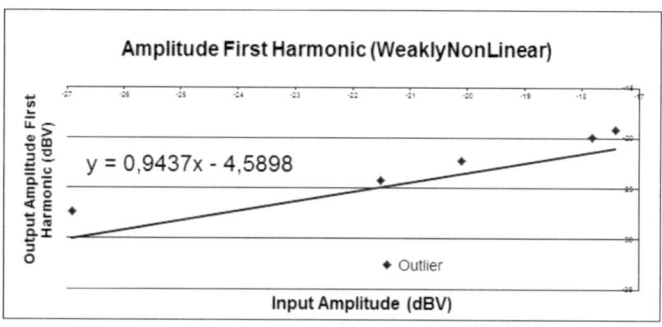

FIGURE 7. The gradient is about 1 and the y-intercept is due to constant term (**??**) and the logarithm of the first coefficient (**??**) divided by the input amplitude.

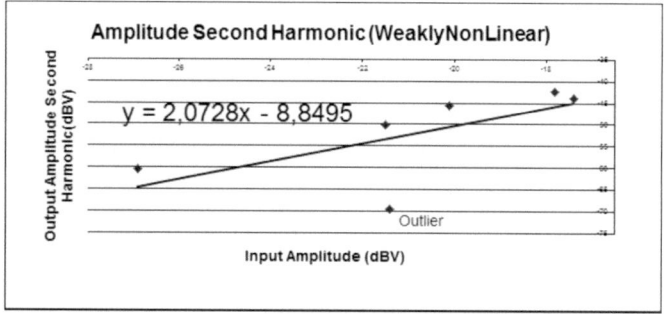

FIGURE 8. As expected the slope is about 2 for the second harmonic. The magnitude of the y-intercept is greater than for the first harmonic as there are more consant terms in the Taylor expansion that have to be taken into consideration.

3.4. **Strong Non-Linearity.** On the function generator, the DC-offset was set
to -2.6V. The respective peak to peak voltages were then measured using the PicoScope. Compared to weak non-linearity we could then also observe the third
harmonic. The following three figures (Figure 9, 10 and 11) show the gradient of
output amplitude versus input amplitude in (dBV), hence they relate to the harmonics. The higher the harmonic the more imprecise the measurements. The signal
strength was reduced and hence manual measurements with the ruler function of
the PicoScope Software got more imprecise.

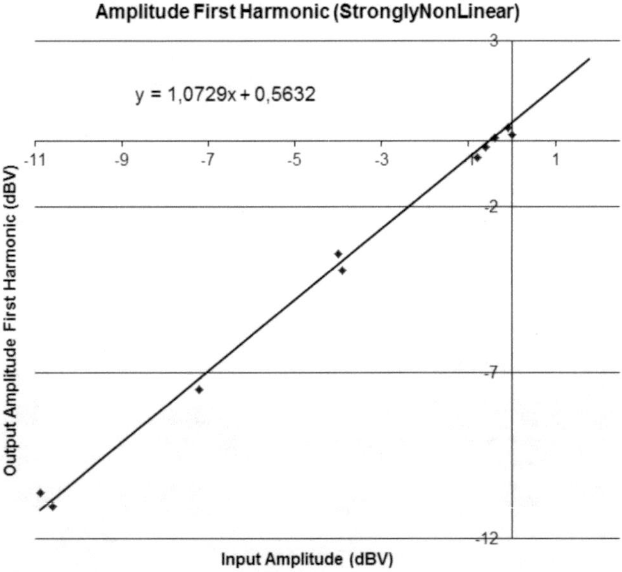

Amplitude First Harmonic (StronglyNonLinear)

$y = 1{,}0729x + 0{,}5632$

FIGURE 9. The gradient is about 1 and the y-intercept is due
to constant term (??) and the logarithm of the first coefficient
(??)divided by the input amplitude. As the measurements took
place in the strong linearity regime, an additional term of $\cos(\omega t)^3$
has to be taken into consideration.

Amplitude Second Harmonic (StronglyNonLinear)

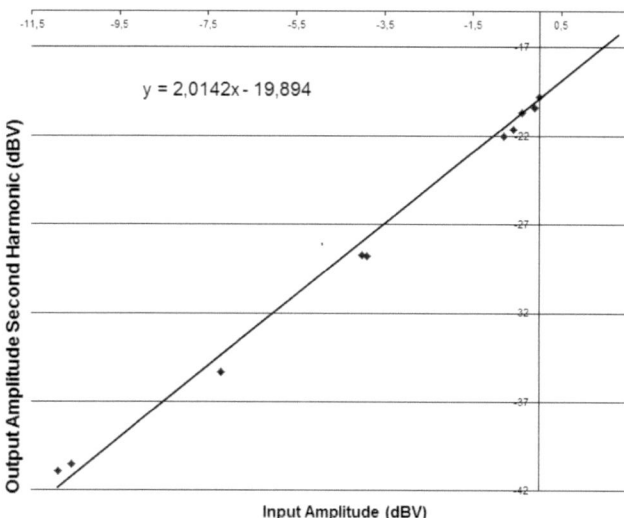

FIGURE 10. As expected the slope is approximately 2 for the second harmonic.

Amplitude Third Harmonic (StronglyNonLinear)

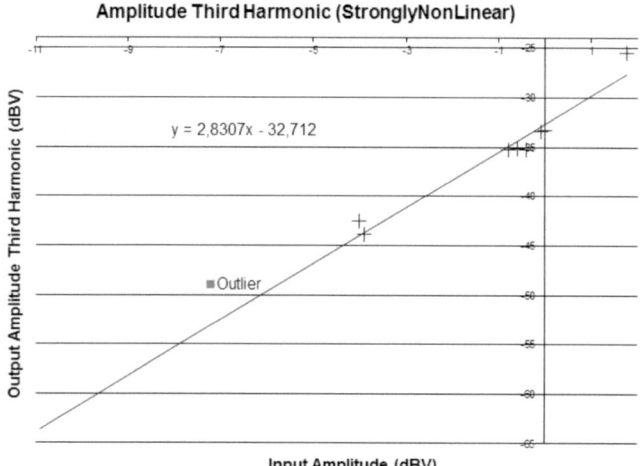

FIGURE 11. As expected the slope is approximately 3 for the third harmonic.

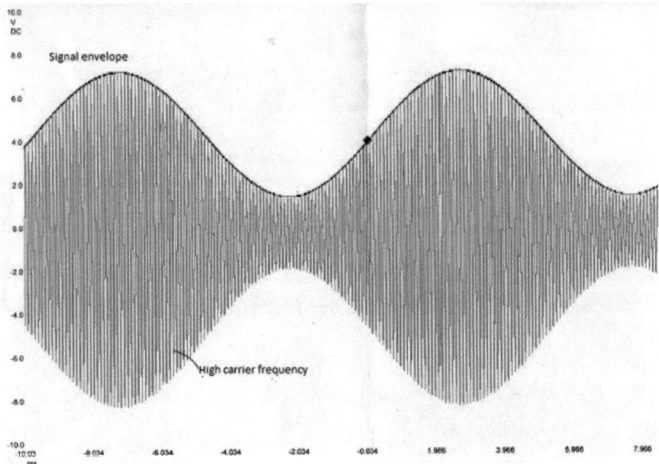

FIGURE 12. The time domain of the signal. The high carrier wave with the signal as its envelope.

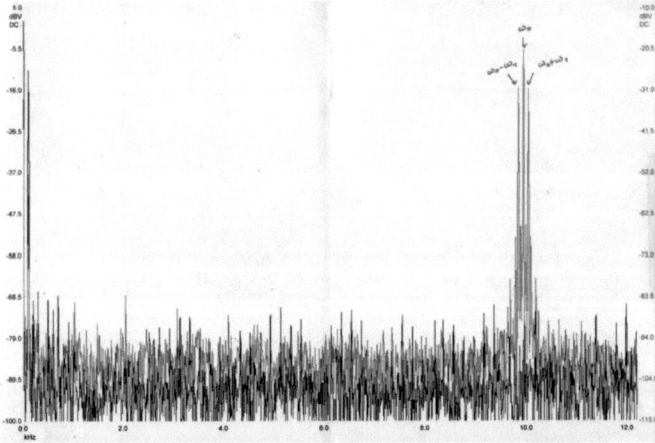

FIGURE 13. The frequency domain of the signal. The original frequency of the signal is visible in the low-frequency area, whereas the two sidebands are next to the carrier frequency. The Signal-to-Noise Ratio (SNR) is 74dB, which is very good for our purposes.

3.6. **Amplitude Demodulation.** The signal was then recovered from the modulated waveform by a further multiplication with the carrier. The amplitude modulated audio signal on channel X and the carrier frequency on channel Y were then connected to the circuit. The transmitted Ethernet output was demodulated and the output was identified using earphones connected to the FILT output(??, which as a low pass filter removed the high frequency carrier. The signal was identified as a UK radio station.

4. Discussion

4.1. Common occurence of outliers. On the one hand, this can be due to imprecise measurement techniques as in Figure **??**. On the other hand, the point labelled outlier in (**??**) changes the slope of the Excel trendline from 2.83 to 2.46, hence it was excluded from the fit. As it is at a very low amplitude, it is not in the strongly non-linear regime anymore, so that it is valid to discount it from the analysis. In general, the errors can be reduced by a more systematic way of measuring the data.

4.2. Potential Errors. We tried to reduce the potential errors by careful design. The circuit board was already built, so that we as experimenters could not introduce any manual setup errors. The AD633 has errors of less than 0.5% [?]. The probes were calibrated several times in order to reduce systematic errors. In general, this experiment was more about the qualititative investigation of non-linear behaviour rather than trying to specify concrete values for the behaviour of some components. Hence, the errors are overall negligible. Please note that our errors are too small to be visible on all our graphs.

4.3. Trendline fitting of logarithmic plots. In general, it is quite problematic to fit a trendline to represent a power-law for log-log plots. [?]In order for them to provide reliable results, they require much more data than it could have possibly been provided in this experimental frame and with this data analysis software. An alternative to identifying power-law probability distributions would be Pareto quantile-quantile plots (or Pareto Q-Q plots). It would also be feasible to apply the maximum likelihood method to our data and use the chi-squared statistic and its distribution function as a measure of goodness of fit. One example of an obvious bad trendline fit that led to a very high R^2 value, precisely 1, is the one of output frequency versus input frequency in the Frequency Doubling experiment (**??**).

5. Conclusion

The theoretical predictions were clearly verified.

(1) This experiment showed that the AD633 multiplier acted indeed as a non-linear element in the circuit.
(2) The effects of different settings of the AD633 were investigated in their frequency as well as amplitude response.
(3) Frequency Doubling was observed when squaring a signal. This is theoretically predicted in **??** and our data agreed fairly well**??**.
(4) Different degrees of non-linearity could be observed by varying the DC level. For a small, sinusoidal variation about a large DC level, the system was found to be weakly non-linear. For a high amplitude and a low DV offset we observed strong non-linearity. Compared to weak non-linearity, we were able to observe the third harmonic as well. This is due to the Taylor expansion of the square-root function of the input voltage. **??**
(5) An amplitude modulated audio signal was multiplied with a carrier frequency to recover the demodulated signal.

6. References

References

[1] Henson, R. N. (2011). "How to discover modules in mind and brain: The curse of nonlinearity, and blessing of neuroimaging. A comment on Sternberg (2011)". Cognitive Neuropsychology, 28 (3,4), 209223.
[2] Donge J.F.;Zou1, Y.; Marwan, N.; Kurths, J. (2009). "Complex networks in climate dynamics". Eur. Phys. J. Special Topics 174, 157179.

[3] Franken, P.; Hill, A.; Peters, C.; Weinreich, G. (1961). "Generation of Optical Harmonics". Physical Review Letters 7 (4): 118.

[4] http://demonstrations.wolfram.com/AmplitudeModulation/ (29/11/2012)

[5] Analog Multiplier Data Sheet
http://www.analog.com/static/imported-files/data_sheets/AD633.pdf (29/11/2012)

[6] Clauset, A.; Shalizi, C.R.; Newnham, M.E.J. (2009). "Power Law Distributions in Empircal Data". Journal SIAM Review. Volume 51 Issue 4, 661-703.

[7] Michaelmas 2012 1B Lab Manual "Systems and Measurement".